HOW TO GROW ROSES

Easy guide to planning, planting and growing a rose flower

Robinson Hud

Copyright © by Robinson Hud 2024. All rights reserved.

Before this document is duplicated or reproduced in any manner, the publisher's consent must be gained. Therefore, the contents within can neither be stored electronically, transferred, nor kept in a database. Neither in Part nor full can the document be copied, scanned, faxed, or retained without approval from the publisher or creator.

TABLE OF CONTENT

Introduction	**4**
Overview of the Rose Plant	6
SECTION 1	**8**
The classifications of roses	8
Old garden roses	8
A damask	9
China rose	10
Gallicas	10
Bourbons	11
Albas	11
Centifolias	12
Modern roses	13
The Hybrid Teas	13
Floribunda	14
The grandifloras	15
Polyanthus	15
Landscape roses	16
The Groundcover Roses	17
Miniature Roses	17
Hybrid Musks	19
English roses	19
The wild roses	20
The planting of roses	**23**
Understanding and recognizing the Six Key Rose Parts	24
How to bury the bud onions	29
How to grow your rose (requirements)	31
Light	32

Water	33
Moisture	34
The soil	35
Fertilizer	37
Pruning/Cutting back:	40
The process of propagation	45
Harvesting	49
Popular varieties that are simple to grow for beginners	**53**
Rosa 'Lady of Shalott	53
Rosa 'Bolero'	54
Rosa 'Charles de Mills '	54
Rosa 'Bonica'	55
Pests and diseases	**56**
Aphids	56
Weevils and mites	57
Japanese beetles	58
Thrips	59
The sawflies	60
Black spot	61
Powdery mildew	63
Rose Rosette disease	64
Questions That Are Frequently Asked	**66**
Do roses pose a threat to animals?	66
Which flowers have the most delicate aroma?	66
From which roses do hips originate?	66
The roses blossom at what time?	67
Conclusion	**68**
SECTION 2	**69**

31 Companion Plants to Grow with Roses This Season 69
About Companion Planting 71
Rose Companion Planting Benefits 73
Top Companions for Roses 73
Companions Highly Recommended 75
Conclusion **142**
SECTION 3 **143**
15 TIPS FOR GROWING BEAUTIFUL ROSES IN POTS OR CONTAINERS **143**
Introduction **144**
Advantages of Planting from Containers 148
1. Selecting the Appropriate Genus of Rose 151
Which kind of roses are most suitable for growing in containers? 152
2. Choosing the right container 158
3. Choose the right location 162
4. Choose the right Soil 164
5. Prepare your rose before planting 165
6. How to Plant the right way 166
7. Fertilizer application 166
8. How to water your plants 170
Conclusion **181**

Introduction

Roses are a classic example of garden beauty. Due to the fact that roses come in a broad variety of colors, shapes, and sizes, there is a rose variety that is suitable for any garden. Roses, which are frequently referred to as the "Queen of all flowers," provide your garden a sense of history, elegance, abundant flowering force, and an unparalleled smell. Throughout this book, I disprove the notion that roses are a fussy flower. You will learn how to care for them during this season and beyond if you continue reading!

You can not go wrong with a rose. The American national flower is the sunflower. It has a long and illustrious history of cultivation that dates back at least 5,000 years, and it has been a part of the gardens of monarchs, queens, and even people who garden in their own homes.

There are 150 different kinds of roses, and there are 30,000 different types to pick from. Roses are graceful and dazzling in color and form.

There is a common perception that roses are challenging to cultivate; however, you should not let this deter you. Roses are just flowering plants, and if you are familiar with some fundamental planting guidelines, they are just as simple to cultivate and maintain as the other plants in your garden collection. The vast majority of contemporary kinds of these well-liked flowering perennials are resistant to both pests and diseases, and they do not require any extra care or attention from you.

On the other hand, just like any other plant, it is beneficial to be familiar with certain fundamental maintenance essentials in order to ensure that your roses remain healthy and continue to bloom.

Continue reading to learn more about growing roses, including how to give them the optimal habitat and how to adequately care for them.

Overview of the Rose Plant

Nature of the Plant: Perennial
In the Rosaceae family
Rosa is the genus.
Geographical regions of origin: Asia, Europe, North America, and Africa Season Spring-Fall
Hardiness Zone 2-11: Period of Time The risk of exposure
From direct sunlight to partial shade: A plant At a distance of three to four feet
Height of the Planting: 16 to 24 inches, 1 to 60 feet in height.
Pests are deep and few in number.
The diseases caused by aphids, beetles, mites, and thrips
Spots of blackness, powdery mildew, and rust
Maintaining: a moderate amount
The soil is loamy and well-drained.
pH (6.5) of the soil is acidic and neutral.

Combining Nepeta, Lavender, and Coneflower with Plants

Bees, butterflies, and birds are all drawn to it.

SECTION 1

The classifications of roses

Throughout history, roses have been revered as a symbol of love and romance, and they have also been utilized for therapeutic purposes.

For the purpose of making it easier to sell and produce a large number of diverse cultivars, the three primary groupings of roses are further subdivided into sixteen popular varieties, each of which is distinguished by its heritage and qualities.

Old garden roses

These roses were already in existence prior to the year 1867. The vast majority of them only bloom once in the spring and have a potent aroma. All things considered, they are extremely hardy and can withstand a wide range of conditions in the garden. Roses that are older can be found in a wide range of colors and sizes, including climbers and rambler kinds.

There are a few varieties that you are likely to come across at your neighborhood garden center:

A damask

There is a lot of recognition for the stunning pink flowers that damasks produce.

Damask roses are a type of rose that bloom in a shade of pink that can range from light to dark. In addition to having an upright habit, they might have a single or double flowering pattern. Shrub roses have the potential to grow to a height of seven feet. The aroma of these flowers is reminiscent of a classic deep rose, and they only bloom once.

China rose

The blooms of this particular rose variety are noted to be multicolored.

The China rose blooms throughout the year in a variety of warm colors, including pink, red, and yellow. Because of its multicolored effect, the cultivar known as "mutabilis" is quite popular. This shrub can grow to a height of 8 to 10 feet.

Gallicas

Gallicas are well-known for the vibrant pink flowers that emerge from their flowers.
The Gallicas are available in a variety of vibrant, deep pinks and reds. Their height ranges from three to four feet, and they have a tendency to form thickets or natural hedges. In the late spring, its bloom lasts for three to four weeks.

Bourbons

There is a wide variety of solutions available for this style, including climbers.
Bourbons are robust flowering shrubs that produce huge flowers that range from pure white to deep crimson in color ranges that are eye-catching. Both stripes can be found on certain cultivars, such as the "Variegata di Bologna" variety. Bourbon is

characterized by a robust and enticing aroma, and it is also quite substantial. There are a lot of climbers that belong to this category.

Albas

When it comes to chilly climates, Alba is an excellent option.

In addition to being disease-resistant and hardy shrubs, albas are an excellent option for cold locations. They are able to be trained to become climbers. They are characterized by big, double blooms that are pink and creamy white in color, and they are complemented by foliage that has a stunning blue tint. Once a year, they bloom for a considerable amount of time.

Centifolias

Centifolias, which are often referred to as cabbage roses, are distinguished by the number of petals that are present in their flowers.

centifolias are distinguished by the high number of petals that they possess. They are often referred to as "cabbage roses" due to the full and rounded shape of their blooms. Typically, they are pink or white in color, and their flowers are so heavy that they hang down from their stems. "one hundred petaled rose" is what the name "centifolia" means. In addition to having a wonderful and fragrant aroma, they only bloom once per season.

Modern roses

At the present time, the most popular varieties of roses among gardeners are modern roses. From the showy Hybrid Teas to the hardy shrub or landscape roses, they are all varieties that are available. Some of the colors that these roses blossom in include red roses, purple roses, pink roses, and many other colors in between. The following is a list of categories that you might come across:

The Hybrid Teas

The hybrid teas are a traditional type of florist rose and are available in a wide range of hues.
a traditional florist rose, characterized by a long, strong stem that supports a shapely bloom. The aroma of hybrid teas can range from mild to robust and delicious, and they are available in a broad variety of colors. Their height ranges from three to eight feet, and they bloom in cycles throughout the season. They are a shrub that is erect.

Floribunda

It is common knowledge that floribundas are recognized for their compact nature.
Floribundas are grown to a height of around three feet and are bushy and compact. When they bloom, they produce a large number of flower clusters. They often bloom throughout the entire season and are hardy up to zone 5. Floribundas are available in a wide range of hues to best suit your needs.

The grandifloras

Grandifloras can reach heights of up to six feet and bloom continuously throughout the summer.

The Grandiflora is the result of combining the bloom strength of Floribundas with the traditional shape of hybrid teas. It is possible for flowers to appear on single stalks or in clusters. They bloom throughout the entire summer and come in a wide range of colors. They reach a height of up to six feet.

Polyanthus

These shrubs are recognized for their compact and resilient nature, which is why they are called polyanthus.

From spring till frost, polyanthus are hardy shrubs that are compact and covered in clusters or sprays of fluffy tiny flowers for the whole growing season. They reach a height of two to four feet. 'The Fairy' is a cultivar that is quite popular.

Landscape roses

Knockout roses are only one of the many varieties of roses that may be used in landscaping.

The landscape rose, also known as the shrub rose, is a vigorous flowering plant that requires minimal care and thrives with little attention. There is a wide range of shapes, forms, and sizes available for these flowers that repeat themselves.

For instance, the Knockout Rose® and Easy Elegance® Series are two examples that are rather well-known. These roses can be seen flowering in a wide variety of colors, including red, pink, white, bi-color, and bright yellow, among others.

The Groundcover Roses

There are a wide variety of colors that groundcover roses bloom in, and they are known for their low-growing characteristics. To put it simply, groundcover roses are a variety of landscape rose that grows at a very low height. They can reach heights of one to three feet and have a habit of spreading out. Many different kinds of groundcovers are available in a wide range of bloom shapes and colors. In most cases, they are resistant to cold and produce several flowers. Among the most well-liked variations are the Flower Carpet ® Series and the Drift ® Series.

Miniature Roses

There are a wide variety of colors that miniature roses bloom in, yet they do not grow to be extremely tall.

Miniature roses are a versatile form of rose that are smaller in size and remain at a height of less than two feet. It is possible to find them in virtually every color and bloom shape imaginable. It is possible to plant them in the ground among your other roses, despite the fact that they are quite small.

Hybrid Musks

The hybrid musk has a distinctive appearance and is tolerant of direct sunlight. On tall canes that can reach up to six feet in height, hybrid musks have flowers that appear to be fragile and are composed of cream, pink, and apricot. In addition, their foliage is lustrous and dark green. The disease resistance of these roses is exceptional, and they are resistant to shade in many cases.

English roses

One of the most well-liked plants in cottage gardens is the English Rose.
David Austin is credited with popularizing English roses, which are characterized by deeply scented, cupped blooms that are packed with petals. These creatures are available in a wide range of hues and dimensions, including climbers. Due to the fact that they have big blossoms, an old-fashioned charm, and a modern

simplicity of maintenance, they are much sought for.

The wild roses

The majority of wild roses have solitary blossoms that are open, and they are simple to care for.
Over one hundred naturally occurring plants that are native to North America, Europe, and Asia are included in the category of wild roses. These roses are also referred to as species roses. The majority are pink, and each one has a single open bloom with five petals.

In addition to being utilized in landscaping, they can be found growing naturally in a wide variety of wild settings without any type of intervention. Because they are so simple to maintain and beneficial to animals, they are becoming increasingly popular in gardens.

Here are some wild roses that you might come across:

"Woodsii": The stems of this hardy rose from the western United States can grow up to five feet tall, and the flowers are a lovely pink color.

'Carolina' is a native plant that is native to eastern America and grows between one and three feet tall. It has delicate brilliant pink blooms that open up to yellow cores.

Canina is a European native that may reach a height of up to 15 feet and has beautiful flowers that range in color from pale pink to white. It is commonly referred to as the "Dog Rose." When autumn arrives, it produces a large number of hips that are brilliant, crimson and attractive.

The planting of roses

When the temperature is between 40 and 60 degrees Fahrenheit, the best time to grow roses is at the beginning of spring or in the fall.

Although it is possible to cultivate roses from the seed that is obtained from hips, the success rate is unpredictable and requires a great deal of time and care. Because the majority of roses that are currently available for purchase are hybridized or grafted, the seedlings of these roses will not produce the same bloom as the parent plant.

Because the majority of gardeners are interested in knowing the type of rose they will receive, we will concentrate on planting roses that are either bare-root or container-planted from nurseries.

Roses are classified as woody perennial plants, which means that they continue to grow the next year after flowering, after which they enter a dormant state. When planting, it is important to be familiar with these six essential plant parts:

Understanding and recognizing the Six Key Rose Parts

1. Root Ball
The mass of roots that is located just below the stem of the rose flower is responsible for the collection and storage of nutrients from the soil.

2. Union of Buds
In the lower part of the stem, the point where the canes emerge from the base of the stem. Roses that have been grafted: near the root, where the grafting began.

3. "Canes"
The branches that are developing from the base of the rose plant are the ones that will eventually form the leaflets and buds of the rose.

4. Flyers and leaflets

There are compound leaves on roses, and they sprout from the canes of the plant. The majority are sold in sets of three or five. They are critical to the process of photosynthesis.

5. Eyes of Bud
This fleshy nub on the canes, which is where new growth is beginning to develop. At the point where the leaflets join to the stems, bud eyes form.

The ideal season to grow roses varies from region to region and location to location. Please refer to the USDA Hardiness Zone Map in order to locate yours. Temperatures between 40 and 60 degrees Fahrenheit are good for planting.

An early spring planting or an autumn planting is ideal for roses. When there is no longer any risk of frost, sow your seeds in the spring. You should give them a period of six to eight weeks before the first frost in the fall so that they can establish themselves before going into winter hibernation.

Choosing the most suitable location for planting is the first step in achieving successful planting. The ideal location is one that receives between six and eight hours of direct sunshine each day.

There is no problem with allowing four hours for types that are particularly designated as being shade tolerant. You should choose a spot that is at least three to four feet away from other large plants and is free from root competition from trees.

To avoid scorching your leaves and flowers, it is best to expose them to the sun in the morning and shade them in the afternoon if you live in a warmer region or have particularly hot summers.

There are two primary choices available to you when making a purchase: bare root or container-grown. Both have potential benefits to offer. Container-grown plants are simpler for novices to cultivate because they do not require any additional preparation prior to planting.

When compared to roses purchased from a local nursery, bare root roses are typically more developed and easier to ship. This is because they are purchased online. Their appearance will be that of a bundle of naked canes and roots when they arrive, and they will be absolutely dormant. Do not be concerned; they will go in a very short amount of time.

Before planting a rose that you have ordered with bare roots, you should give it a thorough soak by soaking the roots in a pail of water for a period of twenty-four hours.
How to get your hole for planting
To ensure that there is sufficient room for their roots to grow, it is important to dig a hole that is rather deep and has a number of fissures and tunnels.

Dig a hole that is six to eight inches deeper and as broad as the container that your plant arrived in. If the plant has bare roots, dig a hole that is six to eight inches beyond the root depth. The phrase "dig an ugly hole" is an important mantra to keep in mind. If so, why?

You want your rose's roots to be able to extend into a hole that is irregularly shaped and has a number of cracks and tunnels rather than a perfect, clean circle where the roots have to work harder to go outward in order to get nutrition.

How to bury the bud onions

First, take the plant from the container with care, then loosen the outer roots, and last, insert the plant in the hole that has been made.
Transform your container so that it is upside down, then carefully remove the rose from it. A little massage can help to release the outer roots from the soil and make them

more accessible. After you have made the hole, place the plant inside of it with the roots facing down and the canes facing up.

In order to ensure that the bud-union is buried in the soil, it is essential to ensure that your rose is planted at a depth that is sufficient to cover the knuckle-like growth at the base. By doing so, wind-rock, which has the potential to collapse the plant, will be avoided. For grafted roses, it will also prevent sucker canes from developing from the rootstock, as well as die-back from temperatures below freezing.

After that, get your soil ready. In order to improve root growth and the plant's ability to absorb nutrients, you should use a product that contains mycorrhizae (fungus) if the soil is poor. Because they have the potential to cause the roots to become charred, bare-root roses should not have any chemicals or fertilizers added to them.

The next step is to backfill the hole with a mixture of organic compost and soil that is equal parts, being sure to cover the roots completely beyond the bud union. Start by giving the dirt a light tamping, and then begin watering it to get it off to a good start.

Last but not least, around the base of the plant with three to four inches of mulch will help to preserve moisture and maintain stable soil temperatures. Some excellent options are straw and cedar mulch.

How to grow your rose (requirements)

You won't have any trouble cultivating roses. As a result of the fact that they were primarily developed by florists and show breeders who were striving for perfection, they have a reputation for being difficult to breed.

There is no difference between garden roses and the other perennial plants in your yard;

they only require the same amount of care as the other plants. Let's take a look at the most important aspects that you will need to plan for, as well as some essential rose care that you can implement to ensure that they develop to their full potential.

Light

In order to flourish, the majority of types require eight hours of direct sunshine.

It has been stated that roses require a minimum of six to eight hours of direct sunshine each day in order to achieve maximum health and bloom output. Certain types are able to thrive in more shady environments.

If this is the case, it will be indicated on their nursery tag or on the information that is supplied on the internet when placing an order. There must be at least four hours of direct sunshine for even shade-tolerant types to thrive.

Water

Once or twice a week, roses require a substantial amount of watering.

It is recommended that you water your plants on a deep level once or twice every week. You should aim to water the plant at a rate of two to three gallons every session, being careful to water the plant at its base. Watering plants from above can result in moist leaves, which can be a breeding ground for diseases such as powdery mildew and black spot.

The installation of a drip irrigation system is a popular choice among gardeners who have vast gardens. In order to get the same level of success, you may give them a thorough soaking by allowing your hose to drip at the base for fifteen to twenty minutes twice a week.

Roses that have just been planted require more water than roses that have been around for a while, so you should plan on watering them three to four times each week.

Moisture

Check the moisture content of your soil on a regular basis to ensure that it is not excessively damp, and make any necessary adjustments. In between watering sessions, roses prefer to let themselves dry out.

The soil

The ideal growing conditions for roses are loamy soil that drains well.
It is true that roses, like other plants, benefit from having nutritious soil. The ideal soil conditions are loamy and well-drained, although they are able to adapt well to a wide variety of soil types, including sand and heavy clay environments.

It is recommended that the soil be somewhat acidic and have a pH that falls between 6 and 7 in order to create the optimal growth environment. Do not be concerned if you are unsure about the soil in your area. Simple tests are available for purchase at your neighborhood garden retailer or can be ordered online. If you want to enhance or decrease the acidity of your body, the findings will tell you the appropriate course of action to take.

Simply working in a thick compost around the soil at the base of the plant is an alternative if you find that to be an excessive amount of trouble. Composts made from organic mushrooms or cow dung that has been aged are always a smart option.

Fertilizer

If you want to ensure that your roses receive the nutrients they need before they wake up, it is advised that you use a granular fertilizer in the spring.

Your roses should be fertilized three times a year: once when they begin to leaf out in the early spring, once after they have produced their first flush of flowers, and once in the middle of summer. This will encourage healthy new growth and more blooms.

Due to the fact that fertilizer can cause the roots of newly planted roses to get charred, it should be avoided. Commence once they have finished the first season of their show. If you want to prevent promoting new growth that will perish in the cold, you should stop feeding your plants six to eight weeks before the date of your final frost.

If you want to feed your plants in the spring, use an organic granular fertilizer that has a gradual release and will continue to supply nutrients throughout the season.

The application of granular fertilizer is straightforward. Follow the instructions for measuring it that are printed on the container, and then sprinkle it on top of the soil that is around your rose while working it in a little bit. Put some water on it, and you'll be done!

Following the first bloom, you may continue to apply the same spring fertilizer, or you can choose to use a fertilizer that is based on seaweed or alfalfa to ensure that your roses continue to thrive. Both liquid and granular forms are available for these. The use of seaweed fertilizer foliarly, which means that it is sprayed directly on the leaves, encourages new growth and may make roses more resistant to infectious diseases.

Applying a second dosage of either fertilizer to your roses in the middle of summer will provide them with the vitality they need for their last fall flush.

Pruning/Cutting back:

It is advisable to trim a rose bush in the early spring, which involves cutting off any sections of the plant that are infected, suffering from death, or both.

The majority of rose pruning should be done in the early spring, when the plants are just starting to wake up and the bud eyes (nubs where new growth begins) are beginning to expand. This is the optimum time as well. The first bloom of the brilliant yellow forsythia bush is a common method that gardeners use to determine when it is the best time to trim the plant. Be patient and wait until the last hard frost has past if you do not have any forsythia in your surroundings.

Be sure to pay special attention to the plant and keep in mind the three Ds. Get rid of everything that is sick, dying, or dead to assist the plant regain its vitality and give it the opportunity to concentrate on new development. Utilizing bypass pruners that are sharp, remove any canes that are dark and lifeless and have not survived the winter season successfully. You should prune the canes back to the point where you can see healthy, green growth if they are half dead.

After that, you should evaluate the general form of the plant. You want to foster a vase-like shape, which has a center that is open and is surrounded by canes that are healthy on the outside. Because of this, airflow is increased, and foliar diseases are decreased. In order to free up the center, you will need to trim back any canes that are thin (smaller than a pencil) or canes that cross and rub against one another.

Always attempt to cut back just above an outward looking bud-eye, which is a bulge on the cane that faces the outside of the plant. This is the best technique to ensure effective pruning. Your new canes will be trained to grow outward rather than into the center if you do this before they are used.

You might think about doing a hard prune on your rose if it has suffered significant damage over the winter or if you wish to regulate its size. The process involves reducing the height of the entire plant by half, or if required, reducing it to a height of five to six inches from the ground. It will in a short time flush out and appear to be in good health.

And last, tidy up! In order to avoid any infections from lingering in your soil, it is important to remove any debris from the area after you have pruned it.

An additional light pruning should be performed on them in the fall in order to bring them into good shape and prevent any illnesses from overwintering before they go into dormancy for the entire winter.

The process of propagation

Selecting a healthy green cane, cutting it with clean pruning scissors, and planting it in the soil are the steps that need to be taken in order to propagate your rose.

The majority of roses are propagated through the use of cuttings. Gardeners who are eager for a challenge are the ones who should try their hand at this exciting adventure, which has varying degrees of success.

At any point in time, you are able to remove a cutting off your rose. While some gardeners like using fragile growth from their plants after the first spring bloom, others prefer to utilize semi-hardwood from their plants from the end of summer to the beginning of fall. Both approaches have the potential to bring about success; nevertheless, I will focus on the semi-hardwood approach here.

To begin, select a cane that is robust, green, and healthy, and that has just recently bloomed. Begin your journey from the highest point of the cane to the point where it starts to become more rigid and less flexible. This is the portion that appeals to you. Cut a part of the cane that is between six and eight inches long using bypass pruners that are sharp. This section should be cut just below a leaf node, which is the point where leaves adhere to the cane.

First, have your soil mixture and container ready. Choose a small container that is between three and four inches in diameter and fill it with potting soil. Ensure that the soil is completely saturated with water.

The following step is to make cuts in the cutting so that new roots may establish themselves. The lowest leaves should be removed, but a few should be left at the top for the purposes of photosynthesis and nutritional absorption. A few inches of the thorns should be removed. By using your fingertips, you may quickly and thoroughly remove the thorns from the plant.

Dip the bottom section of the cutting into rooting hormone powder. This is the area where you have created wounds by removing the thorns during the cutting process. Take your coated cutting and place it in the soil mixture that has been produced. Bury it almost to the bottom of the mixture, leaving the top leaves visible.

Create a little greenhouse for your cuttings as the final step. Remove the cork from the bottom of a big soda bottle made of plastic. For the sake of ventilation, leave the cap off. With the cut side facing down, position the bottle so that it is atop the container that contains your cutting. Because of this, humidity and warmth will be trapped, which will assist your cutting in growing its roots.

Keep the cutting away from direct sunlight and at a temperature between 55 and 75 degrees Fahrenheit. After six to eight weeks, look for the roots! If there are a lot of roots and new growth, you can transplant it to a pot that is somewhat larger than it is now in.

Harvesting

When harvesting, utilize a sharp pruner and do it on a chilly morning.
Roses are not just used to make our gardens look more beautiful. You may use them to beautify and scent your house by bringing them inside.

These are some suggestions that can help you cut the freshest roses possible for indoor enjoyment:

In order to maintain the freshness of your recently cut stems, you should bring a pail or jar of water with you outside. If you want your vase to last as long as possible, collect your flowers on a chilly morning and select buds that are just starting to bloom.

You should locate a set of five leaves that is close to the length you want, and then use sharp bypass pruning shears to cut just above the set. It should be placed immediately into the water bucket you have.

As soon as you get inside, pour cold water into the vase of your choice. You should remove any leaves that will be sitting below the water level, since this might potentially lead to the formation of germs, and then lay your roses into the vase. Every couple of days or anytime the water starts to grow hazy, you should change the water.

Place them in a location that is cool and out of direct sunshine, and take pleasure in them! Depending on the circumstances, they will typically live for around ten days; however, you may extend their vase-life by cutting off the ends of the flowers each time you change their water and by adding a floral preservative.

Deadheading, or the removal of spent flowers, should be done throughout the growing season in order to maximize total bloom output.

Popular varieties that are simple to grow for beginners

Thousands of cultivars are available for selection, giving you plenty of options. If you are unsure about where to begin, some tried-and-true favorites that have won awards are listed below for a fantastic beginning:

Rosa 'Lady of Shalott

There is a lovely shrub known as "Lady of Shalott" that has blooms that are vivid orange in color and with a scent that is somewhat spicy.

It is possible to teach this beautiful plant to grow as a climber or a shrub, depending on the needs of your landscape. The rose in question is a David Austin rose, which is well-known for its traditional blooms that have a great number of petals and modern simplicity of maintenance. 'Lady of Shalott' is a flower that blooms continuously and has a beautiful aroma that is reminiscent of mild spices.

Rosa 'Bolero'

The 'Bolero' variety yields flowers that are delightfully delicate and creamy white in color, and they have a fantastic perfume of tropical fruit.

The Floribunda (cluster blooming) rose known as "Bolero" is a creamy white rose with a highly fragrant aroma. Extremely resistant to illness, it may reach heights of three to four feet. Take this indoors and take pleasure in the aroma of tropical fruits it exudes.

Rosa 'Charles de Mills '

Huge ruffled blooms in lilac, purple, and scarlet colors are produced by the 'Charles de Mills' variety.

An ancient Gallica rose, this rose is the epitome of passionate love. It comes in a variety of colors, including lilac, violet, and scarlet, and it has gigantic flowers with ruffles. Your garden will be filled with a rich, traditional rose aroma when you plant 'Charles de Mills' since it blooms between three and six weeks in the spring or early summer.

Rosa 'Bonica'

"Bonica" is a lovely shrub that has blooms that are fragrant and luxuriant pink in color. This strong shrub is adorned in sprays of flowers that are classic pink in color and fragranced with a mild perfume. The blooms bloom from April till the first frost. The plant known as "Bonica" requires little care and is ideal for hedges.

Pests and diseases

Although roses are hardy and simple to cultivate, they are sensitive to a few illnesses and pests. Roses are very easy to grow. In order to better understand how to deal with the most frequent issues that may arise when cultivating the queen of flowers, let's have a look at some of the most typical difficulties that may arise.

Aphids

Aphids have a preference for sucking the juice from the young shoots of the plant on which they feed.
When spring arrives, aphids are typically the first pests that you see appearing. The fluids of the fragile new growth of roses are among the things that these little, green insects with soft bodies like sucking.

If you come upon them, you shouldn't freak out. In most cases, if you do not resort to chemical intervention, beneficial predators such as lacewings, birds, and ladybugs will eliminate infestations within a week or two. In addition to that, you may use the hose to spray them off. When it comes to gardening, those who don't mind a little bit of yuck factor typically recommend just squishing them with your fingers.

Weevils and mites

When it comes to getting rid of spider mites, the most effective method is to use a powerful spray of water to eliminate them.

Due to their small size, spider mites are difficult to detect. It is possible that you have spider mites if the leaves have become dull and brown, and if they have webs that are sticky and white on the undersides of the leaves. Spider mites can be eliminated with a powerful spray from your hose, much like aphids may be eliminated.

Japanese beetles

In order to manually eliminate Japanese beetles, you can do so by placing them in a container filled with soapy water.
These scarab beetles, which are green and copper in color, originated in Japan but have now become invasive in the United States. During their development as grubs in the soil, they consume rose flowers as well as other types of plants.

Because it has been determined that chemical sprays are useless in managing them, the most effective method of action is to physically remove them. If you begin this endeavor at the first hint of them in the spring, it will be less intimidating. Gather some tweezers and a jar of soapy water for yourself.

Go out first thing in the morning, early in the season, and remove those suckers from your roses by plunging them into the soapy water. Do this as thoroughly as possible.

Destroy the body of the deceased. It is required to repeat the process till the population decreases.

Thrips

Insects with wings known as thrips are incredibly little insects that cause damage to the buds and leaves of plants.

It is possible that you have thrips if you notice that your buds are browning or if they are malformed and never open. The leaves are likewise deformed and discolored as a result of these small flying insects.

The most effective method for dealing with thrips is to cut back any infestations that are apparent and then wait for their predators to come. Eventually, the number of thrips will decrease on its own. In the evening, you may also spray your roses with organic Neem oil; however, you should be aware that this method is harmful to the insects that are useful to your roses.

The sawflies

In the process of feeding on the leaves of the plant, the larvae of this predator leave behind holes and patches that are yellow-brown in color.

Sawfly larvae, which are also known as "Rose Slugs," like feeding on their leaves, leaving behind holes and yellow leaves as a result of their feeding. It is not difficult to identify them since they have the appearance of little green caterpillars. Their damage will be visible to you in the form of tannish blotches and holes on the leaves, which will eventually resemble skeletons of the leaves that they were originally born with.

One more time, when it comes to treating sawflies, little is more. You have the option of picking them off like the Japanese Beetle (the undersides of the leaves should not be overlooked!), or you may spray them with the hose. You may also take a step back and wait for their natural predators to eat them for you. This would be an alternative.

Black spot

This fungal disease leaves the leaves and shoots covered with rusty purple-brown patches. It affects both the leaves and the shoots.

The leaves and canes of the plant are susceptible to the fungal disease known as black spot, which is rather widespread. Symptoms include leaves that are dark and black speckled and have an uneven form. These leaves are frequently encircled by yellow halos. These rusty, purplish-brown splotches are frequently found on canes. It is not a significant problem, however black spots might make your rose less robust.

As part of the treatment, any sick plant material should be removed and disposed of. As spores that flourish in moist settings are responsible for the growth of this fungus, it is imperative that you ensure that your roses have sufficient ventilation and that they dry out completely in between waterings. In order to prevent spores from overwintering in the soil and reinfecting your plants, it is important to clean up properly.

Powdery mildew

Powdery mildew is typically observed on days that are hot and are followed by nights that are chilly and have a high humidity level.
The appearance of this fungal disease on your plant is similar to that of a white powdery covering. There are times when the leaves curl up. It thrives on warm days and chilly nights, and it is exacerbated by conditions that are humid.

In this situation, prevention is the most important factor. Avoid soaking the foliage by watering the base of the plant rather than the top of the plant. Water roses first thing in the morning so that they may completely dry out over the day. Make sure that your roses are protected from the sun.

If you have already been affected by powdery mildew, you should remove all of the areas of the plant that are severely diseased. If such is the case, you might try this natural remedy: combine one tablespoon of baking soda with one gallon of water.

Roses should be sprayed thoroughly, and this should be done once a week or as frequently as required. Additionally, milk spray may be of utility. Make use of one part milk to nine parts water. It is also a prophylactic measure to spray and repeat as necessary.

Rose Rosette disease

In addition to being a red new growth with an abundance of spines, this illness is transmitted by a very small tick.

The Rose Rosette Disease, which is also known as Witches' Broom, is a virus that is transmitted by a very small mite. With an abundance of thorns, it seems to be a fresh growth that is very robust, bushy, and crimson. At this moment, there are no therapies that are effective against Rose Rosette, which will cause your rose to die.

If you are able to establish that Rose Rosette is present, you should take the entire plant together with a portion of the soil that surrounds it, place it in a bag, and then dispose of it. There should not be another rose planted at the same location.

Questions That Are Frequently Asked

Do roses pose a threat to animals?

The inquisitive cat that you have does not pose a threat to rose blooms. No matter if they are a part of a bouquet or a home garden, roses are a flower that may be safely handled by pets.

Which flowers have the most delicate aroma?

The English Rose and the Old Rose are, as a general rule, the roses that have the most fragrant aroma.

From which roses do hips originate?

Hips are produced by every rose plant if the spent flowers are allowed to remain on the plant. On the other hand, certain roses develop hips that are bigger, fuller, and last far longer than others. Many people believe that Rugosa roses have the most delicious hips, and they are also often the biggest and most numerous of all rose varieties.

The roses blossom at what time?

Roses typically bloom between the end of spring and the beginning of October, according to a conventional rule of thumb. There are some that only bloom once, while others bloom continually throughout the spring and summer.

Conclusion

When it comes to providing your garden with elegance, history, perfume, and beauty, there is no other plant that can compare to the rose. It is quite likely that you will not be disappointed in adding roses to your collection, regardless of whether you go for a cheery miniature for your patio or a showstopping climber to impress your neighbors.

Not only do they possess an attraction that is unrivaled, but they are plants just like any other. Through the use of common sense and the following gardening advice, you will soon be able to cultivate the roses of your dreams.

SECTION 2

31 Companion Plants to Grow with Roses This Season

Are you interested in finding the most suitable plants to cultivate with your roses? There is a wide range of plants that not only enhance the appearance of rose gardens but also contribute to their overall health and resilience. I take a look at some of my favorite rose companions, complete with their names, in this chapter.

If you take pleasure in cultivating roses as much as I do, you might be tempted to devote all of your gardening efforts to the addition of additional roses. On the other hand, were you aware that growing roses in a monoculture makes them more prone to disease and pest infestations?

Not only can companion planting make your yard more beautiful, but it also has other benefits. In addition to this, it generates biodiversity, which in turn attracts beneficial insects, drives away unwanted pests, and improves the condition of the soil, all of which are essential for robust roses that blossom brilliantly.

That being said, this year, you should avoid making the rookie mistake of leaving your roses alone. When you want to develop a healthy ecology in your garden, add a few companion plants. Having to spend less time fighting common rose illnesses and pests can result in a more enjoyable gardening experience for you as well.

If you are unsure of how to begin, I have compiled a list of the rose companion plants that I consider to be the most beautiful and beneficial to your garden. I have also included options that compliment your flowers to make them even more spectacular, that is, if you are more concerned with their appearance than with their functionality. Let's get down to business!

About Companion Planting

The process of creating an ecosystem that is mutually beneficial by integrating a number of different plant species in a single area is referred to as companion planting.

Within a natural ecosystem, a wide range of plant species coexist and coexist with one another, working together to give advantages that are mutually beneficial. In the rose garden, companion planting, which is also known as intercropping or polyculture, offers the same benefits by introducing a variety of plant species into a single area. Additionally, it is also known as polyculture.

Ever since the beginning of time, people who grow vegetables have been aware of the advantages of companion planting. It is possible to lessen the amount of toxic sprays, costly fertilizers, and weed killers that you require if you cultivate plants that are mutually beneficial to one another.

As a consequence of this, your plants will be healthier, more resistant to diseases and pests, and your garden will be more attractive! Allow me to take a more in-depth look at the process of companion planting as well as the advantages that it offers.

Rose Companion Planting Benefits

An increase in aeration and drainage, a decrease in the danger of disease, and the ability to attract beneficial insects and repel pests are all benefits that companion plants can provide.
Companion plants are known to improve the overall health of our garden. However, how do they function? Here are some of the ways that your roses will benefit from companion planting that is done intentionally.
Control via Biological Means
Shade/Shelter Made of Living Mulch for the Prevention of Disease

Top Companions for Roses

It is important to take into account both the ornamental and ecological worth of the companion plant you choose.

The ideal partner for your roses is going to be determined by the aims you decide to pursue. Although there are a number of well-documented advantages associated with certain plants, such as alfalfa, you might not like the way that they look when they are mixed in with your roses.

In my opinion, you ought to take into consideration both the ecological worth and the ornamental value. The beauty of roses, in addition to their aroma, is the primary reason for our attraction to them. To our good fortune, there are a wonderful number of choices that offer both.

When planting companion plants, it is important to take into consideration the time of flowering, the nutritional requirements, and the growth habit. You should avoid planting plants that may crowd your roses, cause them to be overshadowed, or compete with them for nourishment. Keeping all of this in mind, I wish to introduce you to some of my most beloved companions, along with a brief summary of the reasons why they should be included in the rose garden.

Companions Highly Recommended

The following plants, for the most part, offer attractive blossoms and are beneficial to gardens. In addition to providing soil cover and enhancing the color and form of your roses, the pollen that is found in flowers attracts predatory insects that feed on these pests that are harmful to roses.

Planting species that are natural to your region will provide the greatest benefits; therefore, I have included some native wildflowers that I enjoy planting with my roses.

1. The agastache

The common name: Hummingbird Mint, also known as Giant Hyssop
Botanical name: Agastache
Plant type: Herbaceous perennials
Bloom colors: Purple
Hardiness zone: 5-10
About the plant

Agastache is a native wildflower that is both beautiful and fragrant, and it has flower spikes that are rather tall. In contrast to the roundness of roses, the erect bloom form is a remarkably beautiful contrast. Purple and blue roses are the most popular variety, and they look stunning when combined with other colors such as white, orange, yellow, and light pink roses.

It is an agastache that pollinators adore. Due to the fact that it is a member of the mint family, it has a potent odor that is effective in warding off deer and confusing rose pests by concealing the aroma of their preferred plants. Its strong anise fragrance might be something you enjoy!

The benefits include effectiveness against pests and deer, attractiveness, and attractiveness.

How to Plant: Agastache grows between two and four feet tall, so you should avoid planting it close to small roses since it will obstruct the sunlight. It is easy to cultivate from seed and thrives in sandy soils that drain well into the ground.

2. Coreopsis

This native plant has a pleasant aroma and draws in a large number of insects that are considered to be helpful.

The common name: Tickseed

Botanical name: Coreopsis

Plant type: Annual and perennial plants

Bloom colors: yellow, yellow

Hardiness zone: 3-9

About the plant

Coreopsis is a flower that is so cheerful and bright yellow in color! If you are looking for a plant that does not take much upkeep and attracts a large number of beneficial insects, this is the one you should get. Coreopsis is a natural plant that is found in North America and flourishes when unattended.

Bees that are local to the area adore these fragrant tiny blossoms, which also have a pleasant aroma. Every year, they will return to zones 4-9 and simply reseed themselves at that time. When planted together, they create a stunning contrast when paired with white or lavender roses.

Benefits: attractiveness and attractiveness and benefits.

How to plant: Plant the seeds in the early spring. They grow in an erect manner and can reach heights of two to four feet. They do not require a great deal of additional water and can thrive in a lot of sunlight.

3. Alliums

Insects and other pests are repelled by the allium plant's lovely, globular, purple blossoms that have a touch of onion aroma.

The common name: ornamental onions
Botanical name: Allium
Type of plant.: Cannabis-like Long-lasting
Blooming colors of bulbs: Purple
Hardiness zones: 4-9
About the plant
Alliums are a part of the onion family and are used for ornamental purposes. They have globes of little purple flowers that are so charming that they remind me of images from Dr. Seuss. These lovely things are not only entertaining but also beneficial to the garden.

Aphids are repulsed by the mild oniony odor imparted by alliums, which humans do not detect with their sense of smell. When the odor penetrates the soil, it serves to discourage Japanese Beetle grubs from settling there. Predatory insects, such as parasitic wasps, are drawn to the nectar of allium plants and will assist you in eliminating other unwanted pests.

The 'Globemaster' variety, which is enormous, and the 'Millenium' kind, which is softer and pinker, are my choices. Perennials can be found in zones 3-8 for both.

Beauty, insect repellant, and helpful attractants are some of the benefits.

How to Plant: Before planting, you should be aware of the full height of the variety you have chosen. Although Globe Master gets to be quite tall, I adore the way it looks when it is interspersed with rose blooms. The smaller types might be grown in front of the climbing roses if you so desired. To cultivate alliums, it is most convenient to use bulbs.

4. Yarrow.
Yarrow contains tap roots that are quite deep, which help to enrich the soil and retain minerals.
The common name: Yarrow
The botanical name : Achillea

Type of plant: Herbaceous perennials
Bloom colors: white and pink
The hardiness zones: 3-9
About the plant

As a result of its abundance of little blooms arranged in a large umbel shape, yarrow is an appealing target for predatory insects such as lacewings. It is also said to function as a nutrient accumulator, since it has deep tap roots that mine the soil for calcium, magnesium, copper, and potassium, which are all essential micronutrients for roses to have in order to produce healthy flowers. When soil is compacted, the extensive tap roots of yarrow can also help break it up.

The delicate leaves of yarrow, which resembles ferns and emits a potent odor, is effective in warding off certain pests. It is available in a number of different colors, but the white version is the one that is most commonly found in the United States. Because it is so easy to spread, you should be ready to remove them from any location where you do not want them to be.

Nutrient accumulator, soil aerator, beneficial attractant, and pest repellant are some of the benefits of this plant.

How to Plant: Yarrow may be grown from seed with relative ease, and a single plant can quickly establish a colony through the use of underground rhizomes. Ensure that you are satisfied with the appearance of the plants before they become a natural part of your garden by planting them in the spring or early summer.

5. *Wild marigolds*

Marigolds are wonderful flowers that are vividly colored and have a perfume that is both a deterrent to pests and an attractant to parasitic wasps and ladybugs.

The common name: Marigold
The botanical name: Tagetes
Type of plant: Annual and perennial plants
Bloom colors: Orange, yellow and red
Hardiness zones 2-11
About the plant

Do not fail to recognize the significance of the unassuming marigold. Marigolds are frequently cultivated with other plants in the vegetable garden as companions. On the other hand, these jovial small blooms in orange, red, and yellow can also contribute to the satisfaction of your roses.

Marigolds, which for a long time were thought to be effective in warding off insects, really operate by emitting a scent that hides the smell of the ornamentals that you wish to protect. For the purpose of preventing aphid infestations and attracting predators such as ladybugs, parasitic wasps, and hoverflies, you should plant them next to your rose crops.

Masking, beneficial attractants, and beauty are some of the benefits.

Marigolds are incredibly simple to cultivate from seed first thing in the spring, and they frequently reseed themselves. Here is how to plant them. Due to the fact that they are essential to the growth of a wide range of plants, I plant them in abundance all throughout my garden. They maintain a low profile, making them an excellent cover for bare soil.

6. Larkspur

The leaves of the annual plant known as larkspur are toxic to Japanese beetles, who are drawn to the leaf of the plant.

Larkspur is the colloquial name, and Delphinium is the botanical name. The plant is a perennial, and its bloom colors are blue and purple.

3-9 are the hardiness zones.

About the plant

A natural method of controlling Japanese beetles is provided by this annual flower. These pests are drawn to the beautiful flowers of the plant, and they consume the foliage, which causes them to become poisoned.

You can use larkspur as a "trap plant" (a crop that pests adore and that serves as a sacrifice to attract them away from your favorite plants) close to your roses if you have a problem with these annoying insects.

There are many different types of larkspur, and some of them generate big spikes of attractive flowers that require staking in order to maintain their stability. The roses that are included in their shape are stunning, and the colors that are available will appeal to any palette. On the other hand, the apricot-colored roses are a wonderful contrast to the strong blues.

Advantages: a plant that attracts Japanese beetles and is beautiful.

Tips for Planting: Larkspur thrives in direct sunlight. The best way to cultivate it is to plant it directly in the garden.

7. Bee Balm

Bee balm is a stunningly beautiful and brilliant bloom that not only offers a range of hues to your flower garden but also attracts insects that are helpful to your garden.

Bee Balm is the popular name for this plant, and its botanical name is Monarda,. Pink, magenta, and purple are the colors of the blooms, and is herbaceous perennial.

4-9 are the hardiness zones.

Bee Balm, which is a member of the mint family, is another stunning plant that can be used to complement roses in the yard by adding vibrant splashes of color. The Monarda Didyma variety, which is one of its variants, is effective against powdery mildew and attracts beneficial insects.

A popular choice for cottage gardens is the bee balm plant. It has the potential to reach a height of three feet and spreads easily. Blooms are available in a variety of colors, including pink, red, white, and purple. A rather ragged appearance can be seen on the flowers. I really like how these are arranged in a row behind a row of miniature roses.

Powdery mildew resistance and beneficial attractant are two perks of this product.

In terms of planting, bee balm is simple to cultivate from seed, and once it has reached maturity, it is simple to maintain. Sowing it directly can be accomplished by distributing seeds on the surface of the soil in the early spring.

8. Astrantia.

Aphids are attracted to the blooms of this perennial plant, which are star-shaped and have a delicately pleasant aroma.

The botanical name for the plant is Astrantia, but its common name is Great Masterwort.Pink, magenta, and purple are the colors produced by this perennial flower. 4-7 are the hardiness zones.

About the plant

The attractive perennial known as astrantia is effective in warding off slugs and snails, and it also serves as a "trap plant" for aphids. In addition, they are resistant to deer, which makes them an excellent option for gardens that are directly exposed to the elements.

The peculiar blooms that they have are available in a wide range of hues and remind me of miniature pyrotechnics. When cultivated at the base of enormous, dense rose bushes that block off sunlight, they will not be discouraged because they prefer a little amount of shadow during their growth.

Trap plant, beauty, and pest repellent are some of the benefits.

How to Plant: They prefer soil that is moist and shade that ranges from partial to full. It is best to plant them in the fall so that they may receive the cold stratification that is necessary for them to germinate. They grow nicely from seed.

9. *Sedum*
In addition to producing blooms that are pale coral pink and have umbellate florets,

these sedums have a long flowering period and are highly drought tolerant.

Stonecrop is the colloquial name, and Sedum is the botanical name. The plant type is perennial Succulent, bloom colors include pink, white, yellow, and red, hardiness zones range from three to nine.

You will adore long-blooming, low-maintenance sedum if you are a gardener who suffers from the same laziness as I do. Plants belonging to this family are characterized by their fleshy appearance, resistance to drought, and the ability to bloom for up to six months in certain regions.

You have access to a vast selection of sedum variations; nevertheless, the kinds 'Autumn Joy' and 'Chocolate Drop' are my personal favorites for growing in conjunction with roses.

The foliage of the 'Autumn Joy' plant is blue-green, and the flowers are a lovely cerise pink color and stand around two feet tall. The plant known as "Chocolate Drop" has blooms that are a light coral pink color and attractive reddish-brown foliage that looks stunning when paired with rose leaves that are a brighter green color.

Both kinds are able to withstand high temperatures and have a tendency to develop clumps, which makes them suitable for use in a rose garden. They offer dense growth that attracts pollinators and acts as a living mulch for the surrounding area.

Advantages include trap beauty, a magnet for pollinators, and live mulch.

To plant, Sedums do not require a great deal of attention. In the early spring, they can be grown from seed or from cuttings, and they are easy to cultivate.

10. Salvia
This perennial herbaceous plant produces flowers that are tubular and dark purple in

color, and they bloom in abundance in the spring.

plant-type: annual, perennial, Sage is the popular name for the plant, and its botanical name is Salvia, the bloom colors are white, burgundy, and turquoise

five to ten (hardiness zones)

Because it looks so good with so many other things, salvia is my absolute favorite plant to use as a filler. Although the deep purples are my favorite, the red blooms of the native Salvia 'Greggii' are also very appealing, particularly when they are planted in the same area as red roses.

Not only do salvias have a stunning appearance, but they are also robust, abundant in nectar, and contain essential oils that have a pungent odor and are effective at warding off pests. The leaves of salvias are velvety and silky, and they feature spikes of tubular flowers. Salvias are a part of the sage family. In addition to annual variants, there are also perennial varieties.

Benefits include attractiveness and protection against pests.

Instructions for Planting: Plant salvia seeds in the early spring. Full sun and soil that drains well are two things that they enjoy.

11. Nasturtiums

The flowers of nasturtiums are colorful and appealing, and the leaves of these plants are large and spherical.

The plant is known by its popular name, Nasturtium, and its botanical name is Tropaeolum majus, Colors of blooms include yellow and orange, and is an annual plant .

2-11 are the hardiness zones.

About the plant

The delicious blossoms and appealing circular leaves of this companion plant, which is one of the most researched companion plants, thrive in soils that are poor. You absolutely must have nasturtiums in your garden because they come in every hue of the rainbow and can behave in either a trailing or compact growth pattern.

Using their ability to attract aphids and squash bugs, nasturtiums might be considered a "trap plant." They attract a large number of insect predators, such as parasitic wasps and ladybugs, due to the abundance of nectar and spicy aroma present in their flowers. In addition to this, they perform the function of a dynamic accumulator, which means that they extract calcium from the soil and make it more accessible to your roses.

In addition to being a good living mulch and weed barrier, nasturtiums grow below the surface of the ground. As an added benefit, the entire plant can be consumed.

Beauty, a trap plant, nutrient accumulation, a helpful attractant, and living mulch are only a few of the benefits received.

Nasturtiums do not require rich soil or a lot of attention when it comes to planting. Put the seeds in the ground a couple of weeks after the last frost, and then walk away from them. Be sure to gather the huge pods that they produce, as they contain ripe seeds, so that you can plant them again the following year.

12. Sweet Alyssum

Honey-scented tiny flowers are produced by this ground cover, which is a wonderful addition to any garden.

Sweet Alyssum is the common name for this plant, and Lobularia maritima is the biological name for it. Colors of blooms include white and purple,and is an annual plant.

The hardiness zones are 5-9.

About the plant

The Sweet Alyssum is a lovely ground cover that grows in neat mounds and produces a dense, densely flowered plant. When planted at the base of roses, it functions very well as a living mulch, and the delicious honey aroma that it emits helps to attract a wide variety of beneficial predatory insects, including the hoverfly, which is responsible for devouring aphids.

A further benefit of this small bloom is that it inhibits the growth of weeds. It looks very stunning when it is permitted to extend beyond the boundaries of the boundary.

Living mulch, attractiveness, and a useful attractant are some of the benefits.

The Sweet Alyssum is a hardy annual that may be planted and will continue to thrive well into the fall. Many times, it will reseed itself. The fact that it grows in such a neat manner makes it simple to fit into any space!

13. Hardy Geraniums

The cup-shaped flowers that this well-liked flower produces are either pink, white, or blue-violet in color respectively.

The popular name for this plant is Sweet Alyssum, and its botanical name is

Lobularia maritima. It is an annual plant, and its flowers are white and purple.

The hardiness zones are 5-9.

About the plant

Hardy Geraniums are able to thrive in tough conditions year after year, do not mind having poor soil, and steadily expand into clumps that give soil cover and save moisture.

One of the hardy geraniums that will go well with your roses is one that has flowers in a variety of colors, including blue, purple, pink, and white. The best part is that their potent aroma is effective at warding off a wide range of unwanted insects, including aphids, mosquitoes, and ticks.

As a result of their preference for partial shade, hardy geraniums are an excellent choice for planting beneath huge rose plants.

Beauty, pest control, and living mulch are some of the benefits.

In order to plant hardy geraniums, it is recommended that they be divided in the fall every few years. If you have a gardening acquaintance, ask them if they have a clump that they are willing to share with you. Aside from that, they are easily propagated through the use of seed or cuttings. It is true that seedlings develop at a very leisurely pace; however, once they reach maturity, they will spread elegantly beneath your roses, concealing their lanky canes!

14. Penstemon

Insects that are beneficial to the plant are drawn to the spectacular tubular blossoms of the penstemon.

The popular name for this plant is Bearded Tongue, and its botanical name is Penstemon. It is a perennial plant, and its bloom colors include pink, gold, yellow, orange, white, and red. Its hardiness zones range from 5-8.

About the plant

Penstemons are native to a significant portion of the United States and are characterized by their spectacular, tube-shaped flowers that are adored by pollinators. They are able to survive in dry conditions, which makes them an excellent partner for established roses that require watering approximately once per week.

Penstemons lure beneficial insects such as parasitic wasps and ladybugs, which in turn reduces the number of aphids that are present in your garden. Not only are they lovely ornamentals, but they are also available in a variety of vibrant colors, including blue, purple, pink, white, and red. Penstemons are hardy perennials, which makes them a choice that requires little to no maintenance.

The benefits include attractiveness.

Instructions for Planting: Penstemons are able to develop quickly from seed and thrive year after year. They can grow anywhere from one to six feet tall, depending on the kind, so make sure to mix them with your roses in the appropriate manner!

15. *Feverfew*

A close-up of the flowering There are a few flowers in a garden that is sunny. The flowers are little and resemble daisies. They have yellow centers and white brilliant inflorescences, and they are gathered in brushes at the very tips of the branches. The greenish-yellow leaves have lobes and are lobed.

Feverfew is an excellent medicinal plant that has a potent citrus aroma that is effective in warding off pests.

The botanical name for feverfew is Tanacetum parthenium, and the plant type is a herbaceous perennial flower.In terms of colors, bloom colors is white, hardiness zones ranging from 5 to 8

About the plant

As a result of the dense row of snowball feverfew that I have planted in front of two of my favorite climbing roses, the aphids that attack them are completely unaffected by the presence of the snowball feverfew. The potent lemony aroma of this medicinal plant, which is frequently used to alleviate migraines, is effective in warding off pests.

Additionally, the cheery white puffs of the feverfew plant are an excellent replacement for roses in floral arrangements. Additionally, the original cultivar, which is also known as "Bride's Buttons," is an excellent alternative companion plant.

It is important to plant them in areas where you do not mind them taking up a significant amount of space. These tiny flowers have a high rate of reseeding and can quickly develop into quick-growing bushes.

Benefits include attractiveness and protection against pests.

In terms of planting, feverfew is a plant that can be grown easily practically anywhere and is easy to cultivate from seed. It is a lovely and healthy combination to plant it close to roses that are taller.

16. "Nepeta"

There are stunning spikes of blue-violet blooms that are produced by this fragrant herb.
Catmint is the common name.
Nepeta is the botanical name.
Plant type: herbaceous perennial, perennial purple is the color of blooms.
3-8 are the hardiness zones.
About the plant
Nepeta, which is often referred to as catmint, is a fragrant herb that is attractive and has spikes of blooms that are a beautiful blue-violet color. "Cat's Pajamas" is my favorite kind, despite the fact that there are many others, because it maintains a low and bushy height.

When roses are interplanted with the colors, the colors seem to be quite stunning. In spite of the fact that it belongs to the mint family, it does not spread very quickly.

The Japanese beetle and aphid are both repelled by nepeta. It can also work with soils that are poor. A perennial that is dependable and a workhorse, nepeta is an essential component of rose gardens.

Benefits include attractiveness and protection against pests.

Depending on the kind that you select, Nepeta can either grow tall and lanky or tiny and tidy. The planting instructions are as follows. I favor the compact, spherical habit of 'Cat's Pajamas,' which can grow right in front of or amongst my roses. It is quite attractive to me. Starting from a seed, it is simple to cultivate, and it will bloom in its first year!

17. "Phlox"

Phlox is a plant that produces flowering plants that are tubular and dense in bulk.
name commonly used: Phlox
Phlox is its botanical name.
Types of plants: annual and perennial
coloration of the flowers: pink, purple, red, and white
3-9 are the hardiness zones.
About the plant
The combination of phlox and roses is very gorgeous, regardless of whether you go for the creeping type to cover your garden in color or the midsize shrub variety in a shade that is complementing. In upright phlox, the flowers are fashioned like stars and are arranged on enormous panicles that are fragrant. The variety that is used for ground cover grows in spreading mounds that are perfect for the front of the border.

You may want to consider planting Phlox Paniculata if you are dedicated to cultivating a native garden. This plant can reach a height of four feet, serves as a host plant for butterflies, and is not bothered by the shadow provided by a vigorous rose bush.

Phlox cultivars that are susceptible to powdery mildew include several. In order to keep the air circulation in good condition, thin and transplant as necessary.

Advantages: It is beautiful and it attracts pollinators.

The easiest way to cultivate phlox is to use either transplants or cuttings as the starting point. In a garden, mature plants frequently produce their own seeds. It is important to evaluate the requirements of the phlox you have picked to ensure that it would thrive in the proper setting. Some types prefer more shade or moisture than others.

18. *The Coneflower*
In addition to its use in medicine, echinacea is also cultivated for its aesthetic value. Coneflower is the common name.

Echinacea angustifolia is the flower's botanical name.

Herbaceous perennial plants are considered to be native plants.

pink, purple, red, white, and orange are the colors of the blooms.

3-8 are the hardiness zones.

About the plant

In addition to having a long blooming time, a deep tap root that breaks up compacted soil, and a long list of helpful insects that like it, the coneflower, also known as Echinacea, is a native plant that is native to North America.

Coneflower has been cultivated for centuries due to its therapeutic properties, but it also adds a great deal of aesthetic value to the garden. The butterfly species that seek out echinacea makes regular visits to the plant.

Coneflowers are not only appealing but also useful because of their bright petals that are somewhat drooping and their huge centers that contrast with the petals. Although the lavender known as 'Echinacea Purpurea' is the most well-known, coneflowers can also be found in a variety of colors, including a vibrant raspberry-red, rusty orange, neon pink, or even multi colored varieties.

Despite the fact that they do not spread rapidly, coneflowers tend to create large clumps. They offer the ideal height for concealing rose canes that are not covered. The coneflower is a trouble-free and dependable plant that may be planted anyplace you require a bright spot.

It is attractive to pollinators and has aesthetic value.

Echinacea is a plant that may be grown from seed, but it must be cold-stratified before it can be planted. If you want flowers next year, plant for the fall.

19. Sea Holly
Sea Holly is a unique flower that has a gray-blue coloration and works nicely in

your perennial garden when combined with roses and other plant species.

It is commonly known as Blue Sea Holly.

Name derived from the plant: Eryngium planum

The type of plant: perennial

Blue, silver, and purple will be the blossom color.

The hardiness zones are: 5-8

About the plant

What is that peculiar flower that is tucked away inside your florist bouquet? It has a steely blue color. Without a doubt, Sea Holly! This sea holly plant is quite fascinating since it has blossom tips that are a brilliant blue color and resemble miniature thistles. Due to the fact that they are somewhat thorny, they are an excellent deer deterrent!

In stark contrast to the delicate, spherical rose flowers, the hard structure of sea holly provides a wonderful contrast. Some variations, such as the eye-catching purple Giant Sea Holly, can grow to be as tall as eight feet, while the majority of cultivars tend to be between two and four feet tall. When planting, be sure to take into account the mature size of your variety, and position cultivars that are taller in the back of the border.

The sea holly attracts a large number of predators and pollinators during the summer months. However, because of their lengthy tap roots, they are difficult to transplant, but they also serve as a soil aerator. Either freshly cut or dried, they make for a stunning display in bouquets.

Discourages rabbits and deer, lessens the amount of soil that is compacted, and acts as a useful attractant.

Planting Instructions: Sea Holly can be grown from seed, but it must be exposed to a cold wintertime environment before it can germinate in the spring. Planting it in the fall yields the greatest fruits. It is recommended that you divide it every few years so that you can have additional plants.

20. Lavender flowers

The perennial garden is a perfect place to plant roses and lavender, which is a classic combination.

the common name: Lavender flowers
"Lavandula" is the botanical name.
classification of plant: herbaceous perennial
coloration of the flowers: blue, lavender, and purple
The hardiness zones are 5-9.
About the plant
If you are perusing a rose catalog and come across a photo of a mixed border, there is a distinct possibility that it contains lavender. The combination of lavender and roses is one that is tried and true — it never fails!

The backdrop for your roses is the ideal example of a cottage garden, with its wispy, flowering stalks of lavender in calming tones of pink and purple. Of course, its aroma is something out of a dream.

In spite of the fact that we take pleasure in the well-known aroma of lavender and even use the herb into our culinary creations, many garden pests find it to be extremely repulsive. There are a number of pests that despise lavender, including fleas, ticks, ants, and snails. However, it is frequently suffocated by bees, which crave the sweet nectar that it produces.

Repels pests, enhances beauty, and attracts beneficial insects and animals.

Planting Instructions: Lavender is a plant that may be grown successfully from cuttings, but it can be challenging to grow from seed. Due to the fact that it does not like soil that is wet, you should let it dry out in between waterings. A full sun is ideal for lavender.

21. Lamb's ear

The leaves of the Lamb's Ear plant are velvety and delicate, and the plant itself is a ground cover.

Commonly known as: the Lamb's Ear

Stachys byzantina is the botanical appellation.

classification of plant: herbaceous perennial

Pink and purple are the flowering colors.

4-9 are the hardiness zones.

About the plant

The tender, velvety petals of the Lamb's Ear flower are begging to be scratched and stroked. It is a great addition to a garden for children. This is an especially stunning ground cover for your rose bed, with foliage that is a vibrant silvery green color that stands out against the darker leaves of the plants that are surrounding it.

Lamb's Ear is a plant that grows low to the ground and gradually spreads into individual mounds. Your roses will not be in competition with it, and it will soften a formal hardscape. Despite the fact that the plant is resistant to the majority of pests, it is occasionally attacked by slugs.

Although the leaves is the primary reason that gardeners cultivate Lamb's Ear, the plant also produces an intriguing purple bloom. Cut them off as they sprout to maintain the appearance of the plant, or you can let them bloom if you prefer the way they look.

Living Mulch and Beauty are also benefits.

How to Plant: Lamb's Ear is a plant that can be divided and transplanted with relative ease. The easiest way to grow it from seed is to do it indoors approximately eight weeks before the last frost. It is a bit of a slow starter initially.

22. *Violas*
There is a herbaceous plant known as violas that produces delightfully charming flowers and thrives in both the sun and the shade. Violets is the common name.

botanically known as the Viola

Types of plants: annual and biennial

the colors of the blooms: blue, purple, and white

1-10 scale of hardiness

About the plant

These small flowers are available in a wide variety of color combinations, and they look just adorable with their adorable little faces. Every day, my daughter, who is five years old, welcomes her violas and relishes the experience of watching the small colonies grow into a miniature ocean of flowers. Their ability to attract pollinators and serve as a living mulch is remarkable.

Violas thrive in both the sun and the shade, and they prefer soil that is somewhat damp but drains well. They are able to bloom for a considerable amount of time and, if allowed to grow a little bit taller, they are a charming addition to bouquets.

One of the most well-known varieties of these adorable little blooms is called

"Johnny Jump Up," and it comes in a colorful yellow, white, and purple blend. A wide range of colors, from cream to practically black and everything in between, are available for violas.

Pollinator Attractant and Beauty Features are Benefits.

Planting violas is simple because they can be grown from seed. Fall is the best time to direct sow in warm areas. If you live in an area that experiences cold winters, you should grow violas in the early spring. They are an excellent choice for use as mulch for roses that are grown in containers.

23. Prairie smoke
Prairie Smoke is a type of wildflower that has seed heads that are exquisitely soft pink and feathery.
Common name: Prairie smoke
Name derived from the plant: Geum triflorum
classification of plant: herbaceous perennial

Colors of the bloom: red and pink

3-8 are the hardiness zones.

About the plant

Prairie Smoke, which was selected as the Plant of the Year by the Garden Club of America in the year 2020, is an option that is both ethereal and delicate for the front of a rose border.

This geum species is indigenous to the northern region of North America, and it is most well-known for the intriguing wispy seed heads that have a delicate pink color. Additionally notable are the pink blooms that nod and the ferny leaves that it possesses.

Prairie smoke contains a significant quantity of pollen and is particularly valuable as a primary source of nectar for queen bees to use when they are initially awakening in the spring. This airy plant can be planted in front of tall pink roses to create a look that is both gentle and romantic.

The benefits include attractiveness and attractiveness.

Planting Instructions: Prairie Smoke is a drought-resistant ground cover that expands slowly and becomes a lovely ground cover. It prefers to be in direct sunlight, therefore you should plant it near roses that do not hinder its exposure to sunlight.

24. A thyme

As a result of its potent aroma, which causes pests to become confused, this herb is an excellent partner for roses.

Name commonly used: thyme

Thymus is the botanical name

The type of plant: perennial

the colors of the blooms: white, pink, and purple

The hardiness zones are 5-9.

About the plant

One of the most excellent companion plants for roses is thyme, which is generally considered to be a flavorful herb. This plant serves as a host for lacewings, which are predators that feed on a variety of garden pests. Because of its powerful aroma, it is also able to conceal the perfume of roses, which confuses pests that are known to attack roses. Thyme is effective at warding off larger pests such as rabbits and deer.

The blossoms of thyme are a delicate shade of purple, while the plant's roots are fibrous and grow deep to loosen thick soils.

As a living mulch, thyme is an extremely valuable plant because it grows low to the ground. Make use of the closely related creeping thyme, which is capable of retaining moisture in dry conditions, to get the illusion of a carpet of color.

The benefits include a fragrance masker, a host plant, and living mulch.

Planting Instructions: Thyme can be grown successfully from cuttings or divisions. It is possible to plant it from seed, however the process of germination can be challenging. A significant number of gardeners begin the process indoors and then transplant it when the weather reaches 70 degrees. The plant is hardy down to zone 4 after it has been established.

25. Lady's mantle

There is a perennial plant known as Lady's Mantle that has huge, brilliant green leaves and flowers that are a fluffy greenish-yellow color.
Commonly known as the Lady's Mantle
Name derived from the plant: Alchemilla mollis
The type of plant: perennial
yellow flowers are the blooming colors.
3-8 are the hardiness zones.
About the plant
In addition to having enormous, scallop-edged, vivid green leaves that dazzle when they absorb rain and morning dew, Lady's Mantle is easily identifiable by its distinctive appearance. Despite the fact that the leaves are lovely enough, the chartreuse

blossoms are what really make this plant stand out.

In addition to being a popular florist filler, Lady's Mantle is a cut flower that lasts for a long time and produces a stunning combination with burgundy and red roses.

The use of Lady's Mantle is a stylish way to frame a border while also attracting pollinators. It is able to tolerate the sun, but it will really like the dappled shade that tall roses provide.

Beautiful, living mulch, and a magnet for pollinators are the benefits.

How to Plant: The Lady's Mantle can be split up and sold to other people. Be patient, since it expands from the seed in a relatively gradual manner. It will eventually expand to the front of the bed after it has reached maturity.

26. Bachelor buttons

Bachelor Buttons are able to thrive in poorly nutrient-rich soil and produce flowers that have a brilliant blue color.

Other common names include Bachelor's Button and Cornflower.

Centaurea cyanus is the botanical name for this plant.

The type of plant: annual

Blue, pink, and white are the blossom hues.

2-11 are the hardiness zones.

About the plant

The Centaurea flower, which is also known as cornflower, blooms in a unique shade of blue that is not found anywhere else. This prolific plant produces flowers that resemble little thistle-like haloes of fringe that are atop miniature thistles. This easy-to-grow plant is stunning when arranged in bouquets, and it comes in two colors: a traditional blue with violet centers and a soft pink and white mixture.

Seeds of these quickly spread throughout your rose garden, so you should only plant them in areas where you want your garden to have a more casual appearance. In point of fact, the variant of Centaurea known as Cyanus is so aggressive in its spread that it has been classified as invasive in several states. Before you plant, you should check with the extension office in your state, or you might use a cultivar such as "Frosty," which has blooms that are pink, white, and blue in hue.

Hoverflies, lacewings, and ladybugs are drawn to the nectar of Bachelor Buttons because of its high sugar content. These insects subsequently feed on the rose pests that they have attracted. They thrive in full light and can survive in soil with little nutrient richness.

The benefits include attractiveness and attractiveness.

How to Plant: This is a plant that is simple to cultivate! Plant seeds straight in the fall in zones 9-10, and in zones 5-8, plant seeds two weeks before the date of your last frost.

27. Black Eyed susans

There are enormous, bright yellow flowers with black cones in the center that are produced by the Black-Eyed Susan.

Black eyed susans is the common name.

Name derived from the plant: Rudbeckia Hirta

Types of plants: biennial and perennial

yellow and gold are the flower colors

3-8 are the hardiness zones.

About the plant

Members of the Aster family, black-eyed Susans are a bright and cheery plant. As a result of their ability to attract pollinators and pest predators, they are suitable for use in any landscaping project. In addition to being a host plant for the lepidoptera moth, they are also attractive to birds, butterflies, and bees, and they will create your garden a more conducive environment for rose varieties to thrive.

Native perennials, such as black-eyed Susans, are hardy and require little in the way of upkeep. Full sun is ideal for them, although they can also survive in partial shade. They produce lovely golden clumps in the border as a result of their self-sowing.

When you want to create a dramatic combination, try pairing black-eyed Susans with roses in burgundy or lavender coloring. In most cases, they reach a height of approximately three feet; however, there are kinds that can grow significantly higher or have a growth pattern that is similar to that of groundcover.

The color 'Goldsturm' is a well-liked choice that retains its color throughout the summer. Specifically, it was honored with the Garden Merit Award from the Royal Horticultural Society.

Benefits include attractiveness and attracting pollinators.

How to Plant: Once the temperature has reached 70 degrees Fahrenheit, directly sow the seeds in the garden in the spring. You can also begin planting them within six weeks of the day that your final frost will occur. Seedlings that have been purchased are successful when transplanted after the risk of frost has passed.

28. Dill.

As a result of its ability to attract certain beneficial predators, the dill is an excellent companion for both your roses and your vegetable garden.

The popular name for this plant is Dill, and its botanical name is Anethum graveolens. It is an annual plant, and its colors are yellow. It is hardy in zones 2-11.

About the plant

It is common practice to cultivate dill in the food garden; nevertheless, it may also be a beneficial neighbor for your roses. As a result of dill's ability to attract predators such as ladybugs, hoverflies, lacewings, and braconid wasps, these insects will feed on the pests that you have. In addition to acting as a trap plant for aphids, it is said to be effective at warding off spider mites.

Dill is characterized by its feathery foliage and its umbel flowers that are bright yellow in color. Although the majority of us think of it as a culinary herb, it is also a lovely garden accent and a filler for bouquets. When combined with more formal garden plants, the airy blossoms and leaves create an attention-grabbing combination.

Enjoy all of the benefits that this garden jewel has to offer by interplanting it with your roses. It is important to let caterpillars that are eating the dill plant to remain because it is also a host plant for the Swallowtail butterfly. In no time, they will transform your yard into an even more stunning location!

Host plant, helpful attractant, trap plant, and attractiveness are all benefits of this plant.

Because this member of the carrot family does not transplant well, the best way to grow it is to direct seed it in the spring after the initial frost has passed. Plants that are tall and lanky are produced by seeds that germinate rather quickly. At the conclusion of the season, allow them to go to seed so that they can be harvested the following year.

29. Garlic

An inflorescence of a garlic blossom is shown in close-up against a background of green that is blurred. Small, white flowers with six petals and conspicuous stamens are clustered together on the inflorescence.

Aphids, snails, and rodents are all pests that garlic kills.

Garlic is the popular name, and its botanical name is Allium sativum. The plant type is an annual bulb, and its bloom colors are white and pink.

4-9 are the hardiness zones.

About the plant

It is widespread knowledge in the field of horticulture that garlic is a powerful plant, despite the fact that it is not commonly used in decorative gardens. One more allium is garlic, which has a strong sulfurous odor that drives away pests such as aphids, slugs, ants, and rats.

Allicin, which is found in garlic, is an antifungal chemical that has been demonstrated to be effective as a "green" fungicide in the garden. Growing garlic in close proximity to crops that are susceptible to fungal disease was found to be more successful than the application of synthetic fungicides, according to a number of studies. If you want your roses to be healthier this season, put them next to roses that are susceptible to powdery mildew and black spots.

It is even believed that garlic can make flowers smell more fragrant, which is quite mysterious. The Herb Society of America recommends engaging in intercropping with garlic in order to cultivate roses that have a more pleasant aroma. Test out this peculiar combination and let me know what you think!

A natural antifungal, a fragrance enhancer, and a pest repellant are some of the benefits.

How to Plant: Remove the individual cloves from the garlic bulb and plant them in holes that are approximately three inches deep. The cloves should be pointed down. It is true that garlic is a perennial plant that will spread slowly in zones 4-9; however, if you would rather harvest it, you should plant it in the fall and then dig it up in the spring.

Aster is a bushy plant that produces a large number of flowers and is characterized by little disc-shaped flowers that are purple in color and have yellow centers.
Aster is the popular name, and Aster is the botanical name. The plant type is perennial, and the bloom color is purple.
4-8 are the hardiness zones.
About the plant
Asters are prized for the stunning fall color they display. In addition, they offer a source of nectar that is essential for pollinators to

replenish their stores before winter arrives. Native species offer the greatest number of advantages to the garden, despite the fact that there are many beautiful cultivars.

Not only do asters provide a welcome burst of color, but they also make excellent companion plants because they continue to bloom well into October and even November, when the majority of the garden begins to fade.

The New England Aster, which is a vibrant purple with yellow cores, and the Aromatic Aster, which is a type that ranges from a pale violet to a sky blue, are two varieties that stand out. They both attract a large number of helpful insects and look stunning when combined with roses.

The attractive mounded plants that are produced by asters have the ability to spread rapidly. When necessary, thin them out in order to preserve their health and shape. If you want to stimulate bushier growth, you should not be afraid to give clumps a good chop in the early spring. Clumps have a propensity to become top-heavy and spread out, so they should be trimmed.

Advantages: Beauty and attractiveness are two advantages.

How to Plant: Growing asters from seed is a fast and simple process. They require cold stratification, so if you want them to bloom in the spring, you should sprinkle them onto garden beds in the fall.

31. Pink Muhly Grass

A late-season perennial herb known as Pink Muhly Grass is responsible for producing a haze of pink clouds that are layered over dark foliage.

The popular name for this plant is Pink Muhly Grass, and its botanical name is Muhlenbergia capillaris. It is a perennial plant, and its bloom color is pink. It is hardy in zones 5-9.

About the plant

Pink Muhly Grass is a breathtakingly helpful plant that is native to the Central and Eastern regions of the United States. It has a gorgeous texture. When planted in large quantities or along a lengthy border, this beautiful grass has a mystical appearance. As a late-season flower, you may plant it with your roses to create a stunning autumn display.

Color pink The delicate plumes of Muhly Grass appear to be a haze of pink clouds floating over the tall, dark green foliage it grows on. The pink flowers are followed by seed heads that are a light tan color and give interest to the winter landscape while also providing food for wildlife.

When planted behind ground cover roses or in front of taller kinds, they provide the nicest appearance because they grow in lovely clumps that are around three to four feet tall.

The Pink Muhly is resistant to deer, does not have any pests or diseases, and can withstand drought once it has established itself. For a dreamy and calming color scheme, plant this easy-care perennial alongside roses that are either pink or cream in color.

Beauty and wildlife are two advantages.

How to Grow: Pink [Plant] It is possible to start Muhly Grass indoors from seed or to directly sow it after the frost has passed. In the full sun, it thrives. At the beginning of spring, remove any stems that are brown or dead.

Conclusion

The use of companion planting is an effective method for enhancing the aesthetic appeal of your garden, warding off diseases and pests, and providing resources for wildlife. Typically utilized in the garden for the purpose of cultivating vegetables, it is now time to apply it to ornamental plants as well. You should select mates that are the most suitable for the color and development pattern of your roses, and then observe the results as they unfold.

During this year, you might discover that having nice plant neighbors makes the process of maintaining your rose garden a little less laborious. You will have the satisfaction of knowing that you have contributed to the survival of endangered bees and butterflies as a reward. Have fun on your roses!

SECTION 3

15 TIPS FOR GROWING BEAUTIFUL ROSES IN POTS OR CONTAINERS

Introduction

Would you be interested in cultivating roses but lack the room to do so in your garden? To make the most of limited space and to increase design flexibility, many roses grow wonderfully in pots and containers. This is a fantastic choice for making the most of the space available. I am going to share some of my best advice with you in this chapter on how to cultivate lovely roses in containers or pots!

Roses in containers are a classic and sophisticated option that may be used to adorn a patio, as a way to enliven a dining area, or as a way to flank an entryway. As long as a few specific parameters are taken into account, the majority of roses are extremely successful when grown in pots.

You are missing out on a wonderful opportunity if you have a container garden but have not yet included roses in your collection. The blooming season of roses is far longer than that of the typical perennial. The majority of rose varieties bloom in flushes beginning in the spring and continuing until the first frost.

Additionally, they grow back year after year, which eliminates the need to purchase annuals every spring, which saves you both time and money. In addition to that, they have a perfume that is quite irresistible. Through the use of pots, you will be able to keep your roses in close proximity to public locations where they can be enjoyed.

When it comes to growing roses in pots as opposed to planting them in the ground, there are a few important considerations to bear in mind. I'm going to share with you my top 15 growing strategies for gorgeous roses that you can use in pots and containers so that you have the highest possible chance of success!

Do you think it would be best to put roses in a container?

There are a few disadvantages to planting in containers that you should take into consideration before you get started. Roses that are grown in containers will require a little bit of additional maintenance in order to thrive, so make sure that you are prepared to devote some time and care to them.

Roses that are grown in containers are more prone to drying out, depleting the nutrients in the soil within a few years, and being more subject to harsh heat, cold, and winds that dry up the soil.

It is possible for roses to remain in a container for an endless amount of time; however, due to the limited space available for their roots to expand, the roses may not achieve their maximum potential size. This is not a problem at all if you choose to maintain your roses in a compact form. Should you intend to use roses to climb up a trellis or arbor, this is something that you should take into consideration.

Every few years, you will need to repot the roses and give the soil a good shake in order to keep up with the growth of larger roses. Despite the fact that this is not a particularly labor-intensive operation, it does constitute an additional garden chore.

In my opinion, the advantages are more significant than the disadvantages. Planting roses in the ground as well as in pots is something I do since I adore the sight of roses blooming and the aromatherapy that they provide when they are on my patio. If you decide that growing roses in containers is the best option for you as well, the following crucial advice will help you get off to a good start.

Advantages of Planting from Containers

A good number of us reside in regions that have soil conditions that are less than optimal. When I dig a hole to plant something in my garden, which is located on top of an old riverbed, my shovel comes into contact with basalt rock. In other areas of my yard, the sand is completely unbroken. Even if you are dealing with conditions that are just as difficult, it is still possible to plant in the ground if you put in the effort. However, planting in containers is a lot less difficult!

Through the use of containers and pots, you have the ability to exercise control over the environment in which your rose is cultivated. You have the ability to select the ideal soil, observe the development of roots from the bottom, and give close attention to the amount of fertilization that is applied. In addition, there is no digging!

In addition to the obvious advantages of maximizing space for gardeners who have tiny yards or an excessive amount of shadow, growing a rose in a pot gives you the opportunity to study it for a period of time before committing to or determining where it will be placed in the garden. It is possible that you may find that the rose does not have the growth pattern or color that you had anticipated, and as a result, you will decide to reevaluate the location that you had planned to plant it.

It is simple and enjoyable to grow roses in containers, regardless of whether you intend to keep your rose in a pot for an endless period of time or simply want to give it a trial period before planting it in the ground. The atmosphere of your entryway, lounging area, or patio can be entirely transformed with only a few containers using this method.

1. Selecting the Appropriate Genus of Rose

When planting roses in containers, it is important to select the appropriate variety of rose.

Despite the fact that climbing roses can be successfully grown in containers, it takes a significant amount of effort to manage them and ensure that they are content. You should not choose roses that prefer to grow to a height of twenty feet or enormous shrubs that want to become giant spreading hedges for your patio pots because these are not the ideal selections.

Any rose that is not suitable for the USDA hardiness zone that you are in should be avoided. There is a greater likelihood that a rose that is too delicate for your climate would suffer in a container than it will in the ground. You should choose a winter-resistant rose variety that is advised for zones 2 lower than yours if you live in an area that experiences harsh winters.

Which kind of roses are most suitable for growing in containers?

You should select roses that will reach a maximum height of five feet or less, as this is the basic rule that you should follow. Extra support could be required for something that is taller. Growing in enormous, heavy pots that you do not intend to move, however, is an exception to this rule.

When the guideline of five feet at maturity is taken into consideration, there are a great many lovely possibilities and different kinds of roses that look absolutely stunning when they are grown in pots.

1. Miniature roses
In spite of their diminutive size, miniature roses have the ability to swiftly fill a container.

There is no question that miniature roses are the best option, and you will be surprised by how quickly they will fill the container that you have chosen with flowers. The songs "Pretty Polly Lavender" and "Candy Sunblaze" are definitely my favorites.

2. The Hybrid Teas

In containers, hybrid teas have the potential to perform well.

To be honest, I think hybrid teas that are packaged in containers appear more appealing than those that are ground. Many of them have a v-shape and canes that are bare-legged, which makes them appear a little strange in the garden but is perfect for a specimen plant that is kept in a container. The bright and cheery perfume of "Julia Child" and the sophisticated and scrumptiously fragrant scent of "Pope John Paul II" are two of my favorites.

3. Groundcover Roses

Another type of hardy rose, ground cover roses are known for their rapid growth and their tendency to overflow the sides of containers.

Roses that are grown as ground covers are an excellent option because they will gracefully unfold out of the container. You can even test them out in baskets that are made to hang! In addition to being a blooming machine, 'Apricot Drift' is a plant that fills pots wonderfully. The 'Flower Carpet Pink Supreme' variety is characterized by its rapid growth and stunning double blooms.

4. Small shrubs

As shown in the picture, an English-style shrub such as 'Roald Dahl' has the potential to create a statement in your yard.

I have a strong preference for bushy, floriferous shrub roses, particularly those with flowers that are reminiscent of the English old-fashioned type. Thankfully, many people enjoy cultivating plants in containers. Experiment with the luminous warmth of "Roald Dahl" or the smoky crimson tone of "Cinco de Mayo."

5. The standards

There is a kind of regular rose known as "Ebb Tide" that is capable of successfully growing in containers.

In order to create the appearance of a miniature rose tree, standard roses have been grafted and pruned. They offer a formal appearance and are especially attractive when they are contained in containers. For the roots and additional height that these will require, pots that are at least 10-15 gallons in capacity are required. Take a look at the basic forms of the absolutely magnificent multicolored rose types known as "Peace" or the velvety and melancholy violet known as "Ebb Tide."

2. Choosing the right container

When searching for a pot, the first quality you should look for is its functionality.
You shouldn't pick a container based solely on how it aesthetically looks. Classic terracotta pots have a beautiful appearance, however they are prone to cracking in the cold winters where I live. The following are some tips that may help you select the most suitable container for your new rose:

1. Size

In most cases, it is preferable to have the largest container that you are able to accommodate. More recent research has disproved the conventional wisdom that you should only relocate plants to a pot that is one size larger than the one they were originally in.

The nursery pots that roses are grown in are typically between one and two gallons in size; however, I always transplant them into five-gallon pots at the very least. This keeps my roses from becoming rootbound and requiring frequent up-potting, which would otherwise be necessary.

In 10-gallon pots, larger roses and standards should be able to thrive for a considerable amount of time. In order to cultivate the largest roses and climbers, you will need to find a permanent place and utilize a container that is either half a barrel or fifteen gallons in capacity. Because these will be too big to relocate, you need to be sure that you are prepared to offer protection over the winter.

2. Type of Pot

There is a possibility that the growth of your plants will be affected by the type of container that you select.

After the size of the pot, drainage should be your major priority. When the earth is wet, roses do not want to sit there. If water is unable to escape, you will create a miniature swamp that will suffocate your rose and result in the roots becoming rotting. Make certain that the container you select has holes in the bottom so that water can drain out, or you can add holes on your own.

3. Pot material

Pots made of terracotta allow roots to breathe and have an enduring appearance. When it comes to growers in regions that are mild, they are wonderful selections. Clay and terracotta are both appealing and long-lasting in a variety of environments; nevertheless, they cause evaporation to occur more quickly and may require more frequent watering.

I have winters that are quite cold, and I discovered the hard way that terracotta is not ideal for the weather. For my purposes, fiberglass, wood, and concrete are all good options. Plastic containers are not as long-lasting as other materials, but they are affordable and hold water effectively.

Plastic planters are an excellent choice for those who reside in hot climates.

As a result, they require less regular watering since they retain moisture. If you want to avoid baking the roots, you should steer clear of dark hues because they retain heat.

3. Choose the right location

Ensure that you put your roses in a region that is suitable for container gardening.

The ability to relocate the plants to whatever location you choose is one of the most advantageous aspects of growing plants in containers. Roses that are grown in containers require a minimum of six hours of bright, direct sunlight, just like any other type of rose. Patios with plenty of sunlight are perfect. In order to prevent light-colored roses from becoming charred on hot afternoons, it is better to plant them in a spot that receives sunlight in the morning and has dappled shade in the afternoon.

Think about the design and growth patterns as well. It is important to provide sufficient space for airflow to shrub roses because they can grow fairly wide and bushy. Those who climb require a trellis or an obelisk in addition to vertical space. You can use roses to scent your meetings by placing them near a bench or patio furniture. Alternatively, you can use roses to greet guests by placing them by the front door. At this point, the fun begins! You are free to relocate your containers if you do not like the way they look.

It is common practice for some gardeners to position their largest potted roses, sometimes known as standards, on the ground directly within their garden. If you choose to do this, you should think about having the bottom of the pot removed. This will allow the roots to grow into the ground over time, which will result in a rose that is healthier and less confined.

4. Choose the right Soil

It is essential for growth that you make certain that you have the appropriate soil.
A potting soil that is rich, light, and has some perlite for drainage is the kind that I love to use. The use of a soilless mix in pots is recommended by some experts; nevertheless, I have found that potting soil is superior and requires fewer amendments.

To begin, select a high-quality organic potting soil and incorporate some compost into it. This combination will supply your rose with a healthy balance as well as an abundance of nutrients to help it adjust to the container it is growing in. The addition of some mycorrhizal fungi will provide an additional boost.

5. Prepare your rose before planting

The steps you take to grow roses will be slightly different if you want to plant them from bare roots.

Depending on whether you got a bare-root rose or a container rose, the process of preparing your rose will appear somewhat different every time.

It is necessary to soak bare-root roses for a minimum of two hours prior to planting them. Bare-root roses are cut bare canes and roots that do not contain any soil. They can be left in a pail of water that is filled up to the crown for a period of up to twelve hours.

If you purchased a nursery rose in a pot, all you need to do is give it a good, thorough watering an hour or two before you transplant it into its new home. It is easier for roses to adjust to new environments when they are adequately hydrated.

6. How to Plant the right way

When planting, make sure the plant is planted at a depth that will cover the bud union.

When you plant your rose, you should make sure that it is planted at a depth that is sufficient to just cover the bud union. The bud union is the knobby region where the canes meet the roots, or the spot where the rose was grafted onto a more robust rootstock. Because of this, the crown will be protected against dieback in conditions of severe warmth, and it will also be less sensitive to destabilization caused by wind.

You should create a mound of dirt that is large enough to cover the crown and bud union, and then water your rose until the soil is completely saturated with water. It is possible that the dirt will settle a little bit at this time, exposing the roots. Until they are completely covered, add more soil.

7. Fertilizer application

Regularly apply fertilizer

When you put your rose in its container, you should make plans to use a fertilizer that has a gradual release.

The vast majority of high-quality potting mixes will already have an adequate amount of slow-release fertilizer, which will ensure that your rose gets off to a good start. Additionally, it is not required to add anything else, and adding an excessive amount of fertilizer as a supplement might actually cause the roots of the rose to become burned.

As an alternative, you should give the rose some time to settle in before applying a liquid seaweed or alfalfa fertilizer.

This should be done after the rose starts to develop new leaves, as well as after each flush of blooms (which occurs approximately three times a year). A minimum of six weeks before the date of your last frost, you should stop fertilizing.

The purpose of this is to prevent the emergence of new, sensitive growth that will perish in the cold and to urge your rose to enter a dormant state for the winter. Following the application of slow-release rose fertilizer in the spring of the following year, the rose should be allowed to spend a season in its container.

If your rose does not produce a large number of blooms, it is possible that the fertilizer is being flushed out too rapidly. This is because container roses require more frequent watering than other types of roses. If further fertilizing is required, you might want to think about using organic foliar spray.

8. How to water your plants

Stick to a water schedule

When compared to plants grown in the ground, those grown in containers will demand a somewhat higher amount of water. It might be a bit of a delicate ballet to water roses that are contained in containers. Insufficient water causes the roots to wilt and become weak, while much water causes the roots to become rotting and soggy. It is only necessary to water roses that are planted in the ground once or twice per week, but roses that are grown in containers will drain more quickly and will heat up more quickly.

Regrettably, the amount of water that is required is contingent upon factors such as the temperature, humidity, soil, as well as the species and dimensions of the container. As a result, there is no norm that must be adhered to blindly. Rather than that, you should make it a habit to check the soil for dryness every other day.

Perform the knuckle test on a rose to assess whether or not it requires watering. Putting your finger into the ground knuckle deep is all that is required. Your rose will be healthy if the soil with that degree of moisture is humid.

Please check back in a few days. Ensure that it receives a complete watering if it is dry. You'll soon have a better understanding of how frequently you need to water your plants. During periods of higher temperatures, it could be as much as every single day!

When watering, instead than focusing on the leaves, you should aim for the base of the roots. Wet leaves, which are sensitive to fungal infections and mildew, are prevented from occurring as a result of this. It is necessary to water the soil until it is damp but not drenched to the point where water can easily drain from the bottom of the container.

11. Prune only when needed

Roses that are grown in containers do not typically require as much trimming as their counterparts that are cultivated in the ground since they tend to develop more slowly. Furthermore, if you want to keep your rose bush in good health, you should remove any damaged or discolored foliage as soon as possible, before it may continue to spread.

You can improve the flow of air through the rose by cutting back any canes that are growing toward the interior of the plant, or you can cross and rub other canes. When you prune roses, your bypass shears should be clean, and you should use rubbing alcohol to disinfect the space between each plant.

Cut back to an outward-facing leaf node, which is a little, elevated nub where new growth begins. This will encourage new development and give the resulting form an attractive appearance.

In general, the greatest time to prune a rose for anything other than disease is in the early spring, when the rose is just starting to push out new leaves. When the weather is moderate, it is possible to undertake some maintenance in the early autumn.

12. Repot and Refresh soil regularly
Keeping the soil in good condition by repotting will result in plants that are healthier.
It is inevitable that your rose will ultimately exceed its new habitat, unless you started with a very large container, which is something I strongly encourage. Indicators that your rose is ready for a new container include the fact that it dries out rapidly, that it need continuous watering in order to maintain its moisture level, that it becomes top-heavy, or that it is rootbound (the roots are compacted and swirl about in the shape of the container, rather than spreading out since there is not enough room).

You are able to lift up containers that are light enough in order to determine whether or not your rose is rootbound. Roots that are protruding through the drainage holes indicate that it is ready to be moved to a larger place. It may be necessary to carefully remove the rose from its container in some circumstances in order to observe the growth of roots.

Root-bound roses are susceptible to girdling and may eventually perish if they are not relocated. Therefore, it is important to check for this condition whenever your rose appears stressed, or at least once every two years, depending on the size of the pot.

This is an excellent time to revitalize the soil of your roses, especially if you need to relocate them to a larger container. After a few years, depending on the size of the rose, the nutrients in the original soil will be depleted, and the rose will benefit from having fresh soil and compost added to it. Compost and fresh soil should be added to the pots on a regular basis, and the earth should be gently mixed through the first few layers. Containers should not be moved.

13. Treat pest and disease immediately

If you become aware of a disease or pest, it is essential to take prompt action against both species.

There are several types of roses that are bred to be resistant to disease, and the majority of roses are resilient. Having said that, there are a few illnesses and pests that certain roses may be prone to because of their particular characteristics. Powdery mildew and other fungal diseases can thrive in environments with high temperatures and high humidity, therefore it is important to keep a close eye on your roses at all times, but especially during times when these conditions worsen.

There is also a disease known as rose black spot that you will need to be on the lookout for. This condition is difficult to treat, and if it is not treated, it can cause damage to surrounding tissue.

When it comes to pests, aphids and spider mites are likely to be the most troublesome offenders for you. If your roses get afflicted with pests when they are planted in containers, you are fortunate enough to be able to relocate the container and maintain their separation from other garden plants while you cure them.

14. Grow some companions plants
Insects can be repelled and plant health can be improved through the use of companion planting.
Roses grown in containers can also reap the benefits of companion planting. A living mulch that retains moisture, stabilizes soil temperature, and can either attract or repel pests is created by underplanting your container rose with plants that have smaller height. This not only makes your rose seem more attractive and hides barren canes, but it also functions as a living mulch.

Take into consideration the spreading and miniature sweet alyssum as an alternative that is sure to please a large number of people and looks stunning with any variety. Sweet alyssum, which is densely blossoming and has a honey-like aroma, is attractive to beneficial predators such as the hover fly, which consumes aphids.

Add some creeping rosemary to your garden if you appreciate the way plants drape over the ground. Not only will it look beautiful, but it will also avoid mosquitoes.

A groundcover rose can be planted behind a stunning thriller such as Pink Muhly Grass, which allows you to take a more daring approach to the planting of the rose. In order to prevent your rose from being suffocated, consider dividing the grass as required.

15. Overwinter correctly

By bringing container-grown plants inside, you may shield your roses from the snow and protect them accordingly.

Since I was just starting out as a gardener, I was unable to offer sufficient shelter for my patio roses throughout the winter, which resulted in their loss. Potted roses do not have the same level of insulation as roses that are planted in the ground, which can cause the roots to freeze. This can make it difficult for potted roses to survive the winter. Extreme winds are another potential threat, and when exposed to the elements, canes can become brittle and dry up.

To ensure that your rose's roots receive the best possible protection, select large containers that contain a greater quantity of dirt that acts as an insulator. It is important to allow space at the top for winter mulch, such as shredded leaves or wood chips that have been mounded, as this will help to retain moisture and preserve the crown.

You should relocate roses to a shelter, such as an unheated garage or shed, if you live in an area with terrible weather conditions. Despite the fact that they are not subjected to precipitation, you need to continue watering them every couple of weeks; otherwise, they will perish as a result of drying out.

Avoid bringing them inside the house at all costs. Because of the heat, they will not be able to enter a state of dormancy. Roses take a break throughout the winter months in order to recharge their batteries and prepare for the next year's blooming season.

If you would rather leave container roses outside, put the pots in close proximity to one another to boost the temperature, and wrap them all in hessian or black plastic bags to provide insulation. Position them in a location that will provide them with protection from the wind.

Approximately eight weeks before the date when you anticipate your last frost, remove winter protection. If it appears that there will be significant shifts in the cycle of freezing and thawing, you should keep your potted roses protected until the temperatures return to their normal range.

Conclusion

If you have been hesitant to cultivate roses due to a lack of space or optimal growth circumstances, I hope that you would give some thought to experimenting with growing them in containers or pots. Container gardening is not only simple and convenient, but it also provides a great deal of flexibility.

Roses in containers not only add a splash of color, greenery, and fragrance to your outdoor living space, but they also return year after year if they are properly cared for at all times. Use these guidelines to ensure that your roses produce beautiful and healthy blooms, and enjoy them!

www.ingramcontent.com/pod-product-compliance
Lightning Source LLC
Chambersburg PA
CBHW071924210526
45479CB00002B/540